MARTIAL'S MAMMALS

Welcome to the fascinating world of mammals found in Namibia.

Heartstone Publishing

Editing: Mark Paxton

ISBN 978-99916-816-3-4

Published by:

Projects & Promotions
P.O. Box 656, Omaruru, Namibia
Tel: +264-64-570283
proprom.monica@gmail.com

Heartstone Publishing
S10 3ET, United Kingdom
www.heart-stone-house.com
bryony@heart-stone-house.com

CONTENTS

Hoofed Mammals

Damara Dik-Dik 2
Steenbok 4
Springbok 6
Klipspringer 8
Black-faced Impala 10
Oryx 12
Red Hartebeest 14
Greater Kudu 16
Blue Wildebeest 18
Burchell's Zebra 20
Hartmann's Zebra 22
Giraffe 24
Warthog 26

Large Mammals

African Elephant 28
Black Rhino 30
White Rhino 32
Hippopotamus 34

Small Mammals

Rock Hyrax 36
Ground Squirrel 38
Meerkat 40
Banded Mongoose 42
Yellow Mongoose 44
Pangolin 46

Felines

Caracal 48
Cheetah 50
Leopard 52
Lion 54

Canines

African Wild Dog 56
Bat-Eared Fox 58
Black-Backed Jackal 60

Hyaenas

Brown Hyaena 62
Spotted Hyaena 64

Primates

Chacma Baboon 66
Vervet Monkey 68

DAMARA DIK-DIK

The Damara dik-dik is the tiniest antelope in Namibia. Because of its small size and good camouflage, it is difficult to spot in its natural environment.

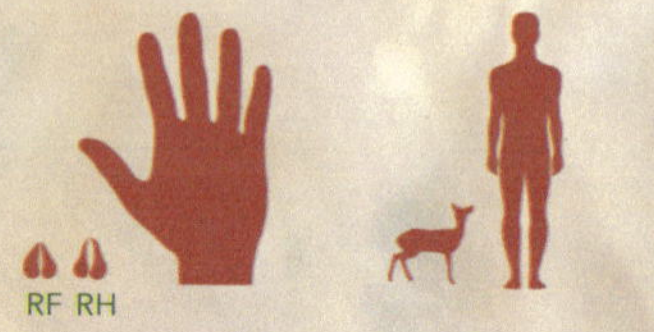

Appearance

Damara dik-diks are grey-brown in colour, with lighter bellies and legs. There is a tuft on their forehead and white rings around their eyes. Dik-dik females are larger than males. Only males have horns.

The horns of the male slant backwards.

Dik-diks can stand on their back legs to reach good food growing above.

Adaptations

Dik-diks are adapted to cope with high temperatures of up to 40°C. Muscles in their nose pump blood through and cool it down. This cooled blood is then circulated through the body, lowering the temperature.

Survival Tactics

Dik-diks let off a sharp whistle when startled and run in stiff-legged jumps. They have good eyesight.

QUICK FACTS

Scientific Name: *Madoqua kirkii*
Conservation Status: Protected
Lifespan: 10 years
Predators: leopard, lion, caracal, African wild dog, python, large raptors
Habitat: dry bush, wooded areas
Speed: 42 km/h
Horn length: 8 cm

Dik-diks get their name from the alarm call they make.

Behaviour

Damara dik-diks are territorial. A male and female normally share a territory. They are seen alone or in pairs. It is believed that they mate for life. Dik-diks mark their territory with urine, dung and a fluid produced by a black gland near the eyes. Males are dominant and cover the female's scent markings with their own.

Conservation and Threat

The Damara dik-dik is listed as Protected. They are hunted for their small bones, which are used in traditional jewellery. Gloves are often made from their skin. Due to the belief that dik-diks mate for life, when one partner dies there is a concern that the living partner will not find another mate, and therefore will not breed again.

The dik-dik has a prehensile nose, meaning it can grasp onto things with it.

© Diganta Sarma

Diet

Damara dik-diks browse on leaves, but also eat shoots, fruit, flowers and pods. They do not normally eat grass. Dik-diks do not need to drink water, but will drink when it is available.

Raising Calves

Female dik-diks are pregnant for 4 months and give birth to a single calf. They are normally pregnant twice a year, giving birth at the beginning and end of the rainy season. The mother suckles her calf for 6 weeks. Males do not help with raising the young. Dik-dik calves reach their adult size by 7 months. At this age, they are forced to leave their parents' territory.

Damara dik-diks are active in the early morning and late afternoon.

STEENBOK

The steenbok is a small antelope found in southern and eastern Africa. It is also called 'steinbuck' or 'steinbok'. They are fairly common.

If a steenbok stands still, it is almost invisible in the bush.

Appearance

Steenboks are golden-brown in colour. Their chins, throats and bellies are white. Inside their large ears is a black leaf pattern. They have black skin around their eyes, surrounded by light fur. Their tail is short and thin. Only males have horns.

Female steenboks have no horns.

Male steenboks have smooth, straight, sharp horns.

Hoof size **Body size**

Steenboks are the only antelope which dig a scrape, and then urinate or defecate in the scrape.

QUICK FACTS

Scientific Name: *Raphicerus campestris*
Conservation Status: Least concern
Lifespan: 6-12 years
Predators: leopard, caracal, African wild dog, cheetah, hyaena, python, large raptors
Habitat: grassland, savannah, woodland
Horn length: 7-19 cm

Diet

Steenboks are herbivores, which browse on leaves, flowers, fruit and shoots. They can survive long periods without water. During dry seasons, they use their sharp hooves to dig up bulbs and tubers.

Raising Calves

Steenboks do not have a strict breeding season. Females are pregnant for 5-6 months. Normally a single calf is born. Calves can walk within 5 minutes of birth.
The calf is hidden for the first 2 weeks of its life. During this time, the mother visits her calf to suckle and clean it. Mother steenboks defend their hidden calves with fierce aggression when a threat approaches too closely. The calf is suckled for 3 months and is considered to be an adult at 7-9 months old.

Conservation and Threat

The steenbok is listed as Least Concern.

Survival Tactics

Steenboks have good eyesight and hearing, and this helps them to spot danger. If a predator is seen, they hide by crouching on the ground. If the predator comes nearer, they flee in a zig-zag pattern, constantly looking for places to hide. They sometimes crouch on the ground where there is good cover and are also known to hide in aardvark burrows.

Behaviour

Steenboks are territorial. They live alone, although a male and female are normally seen together during the breeding season. Territories are marked with strategically placed dung heaps. Males also mark their territories with a scent gland. They are often aggressive when defending their territory. Steenboks are active in the day, but find shade when it gets too hot.

The name 'steenbok' comes from the Dutch word 'steen', which means 'brick'. It refers to the colour of this antelope.

KLIPSPRINGER

The klipspringer is a small antelope found in the rocky areas of Namibia.

Appearance

The klipspringer has a thick yellowish-grey to reddish coat, which has brittle hollow hairs. This cushions the klipspringer and helps to protect it from bruising during falls. This small, sturdy antelope has large hindquarters and closely set hooves, which help it to balance when jumping from rock to rock. It has a brown forehead and a black gland near the eyes. Their rounded ears have a black pattern inside. Only males have horns which are short and spiky and grow out of the skull at a 90 degree angle. Females are slightly larger and heavier than males.

Diet

Klipspringers are browsers which eat plants, fruit, green shoots and flowers. Because they eat many succulent plants, they are not dependent on water.

'Klipspringer' is an Afrikaans word which means 'rock jumper'.

Survival Tactics

Klipspringers are very alert and constantly watch for predators. When threatened, they flee to higher ground, while giving whistling alarm calls.

Adaptations

Klipspringers' small hooves are blunt and rounded, which gives them excellent balance. They walk on the tip of their hooves, which helps them to grip rocks, climb and jump.

QUICK FACTS

Scientific Name: *Oreotragus oreotragus*
Conservation Status: Least Concern
Lifespan: 6-15 years
Predators: leopard, jackal, spotted hyaena
Habitat: rocky areas
Speed: 50 km/h
Horn length: 8-10 cm

Klipspringers are difficult to see in their natural habitat as they stand very still and are well camouflaged.

Behaviour

Klipspringers are active both day and night, but especially during early mornings and late afternoons. They are found alone, in pairs or in small family groups. Males and females form lasting bonds and both defend their territory. They greet each other by rubbing cheeks.

When fleeing from danger, the female normally runs first.

Conservation and Threat

The klipspringer is listed as Least Concern. Although it is hunted for its meat, leather and hide, there are no major threats to its survival, possibly because it inhabits inaccessible areas.

Raising Calves

Klipspringers are seasonal breeders, usually giving birth to a single calf in spring, after a pregnancy of about 6 months. Calves are hidden for up to 3 months and are suckled up to 4 times a day. They are weaned after 4-5 months. Male calves leave their mothers when they are 6 months old, and females leave at 1 year.

Males are protective over their young and keep a lookout for predators.

BLACK-FACED IMPALA

The black-faced impala is a sub-species of the common impala.

Appearance

Impala have a reddish-brown coat, which is darker on the back. They have a black stripe on their tail and two black stripes at the top of their back legs. Impala have black tufts just above their back ankles and white patches above the eyes. Their ears are large with black tips. The black-faced impala has a black stripe on its face and is darker in colour than the common impala. Only males have horns.

Common impala have no black stripe on their face.

Survival Tactics

Impala are always alert. If they see something suspicious, they make a bark-like sound. When attacked, the herd scatters in all directions with a bouncing gait. They weave through narrow gaps between bushes and trees, and jump 2-3 meters high, covering a ground distance of up to 11 meters.

Horns of the male have a lyre shape.

Female impala have no horns.

A territorial male shows off by curving his neck and raising his tail to display the white underside, patrolling his ground, roaring and scent-marking.

Black-faced impala are only found in Namibia and Angola.

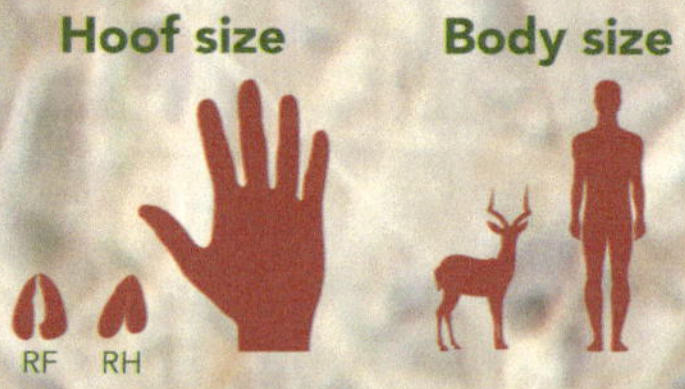

QUICK FACTS

Scientific Name: *Aepyceros melampus petersi*
Conservation Status: Vulnerable
Lifespan: 8-13 years
Predators: lion, African wild dog, cheetah, crocodile, leopard, hyaena and python
Habitat: savannah, grassland, open woodland
Speed: 48 km/h
Horn length: 45-92 cm

Behaviour

Black-faced impala are commonly found in herds of 3-15, although they do occur in larger herds. The common impala can be found in herds of over 100 animals. The size of the herd is also dependent on the amount of water and grazing available. Impala live in 2 types of herds for most of the year – a male herd and a breeding herd. The breeding herd contains adult females, young males and females, and one dominant adult male. During the breeding season, young calves form nursery herds with some attending females.

Raising Calves

Impala have a very distinct breeding season and give birth in December/January. A single calf is born after a 7 month pregnancy. If conditions are harsh, a mother can delay birth for up to a month. The female leaves the herd to give birth in an isolated spot. Most calves are born around mid-day. The newborn is able to walk strongly if necessary but is usually hidden for the first few days.
The mother suckles her calf for up to 6 months.

Half of newborn calves are killed within the first few weeks of life.

Conservation and Threat

The black-faced impala is Vulnerable and nearly extinct in Angola. In Namibia, the population has been affected by drought, hunting and translocation to farms.
To help protect this species, 310 black-faced impala were moved into Etosha National Park during 1968-1971. This helped to increase their population, and there are now more than 1 500 individuals in Etosha. Black-faced impala are also threatened because they can interbreed with the common impala.

Diet

Impala normally graze, but will also eat fruit, herbs, shrubs, flowers, bark and pods. They are water dependent.

SPRINGBOK

The springbok gets its name from its habit of jumping.

Appearance

The springbok is a medium-sized antelope with thin long legs and neck. They have a light yellowish-brown coat and pale underside, separated by a darker brown stripe. Their face is white with dark strips which run from the eyes to the mouth. Their ears are long and pointed and their short tail is white. They have an area of long hairs for display purposes from the middle of their back to their tail. If a springbok is excited or alarmed, this patch opens and shows a crest of white hair. Both males and females have black, curved horns.

The springbok is a common antelope in Namibia. They are adaptable and live in most areas.

Hoof size

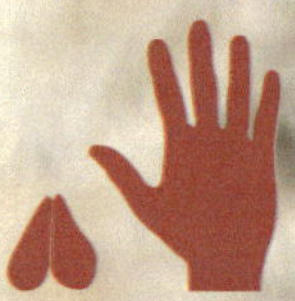

Body size

Behaviour

Bachelor males and females form separate herds. Territorial males usually keep female herds in their area. In the breeding season, non-dominant males form bachelor groups.

QUICK FACTS

Scientific Name: *Antidorcas marsupialis*
Conservation Status: Least concern
Lifespan: 10 years
Predators: lion, leopard, cheetah, hyaena, African wild dog, black-backed jackal, caracal
Habitat: savannah, semi-arid areas
Speed: 88 km/h
Horn length: 36-48 cm

Survival Tactics

Springboks often go into bouts of repeated stiff-legged high leaps of up to 2 metres into the air. This is known as 'pronking'. Springboks pronk when chased by predators, but also in times of joy with the first rains. The white fan is lifted on the back, which emits a strong scent of sweat.

Raising Calves

Springboks give birth all year round, but mostly when there is plenty of food available. A female is pregnant for 168 days and usually gives birth to only one calf. The young calf is born under a bush or other shelter. The mother returns to suckle it several times a day. After a few days, it joins the herd. Mothers and calves sometimes form a nursery herd. Springbok calves are weaned at about 6 months old. Males leave to join a bachelor herd. Female calves stay with their mother's herd.

Diet

Springboks are mostly grazers and prefer short grass. During dry times they eat shrubs, flowers, leaves, fruit and seed pods.
They get most of the water they need from their food, and can survive many days without drinking.

Conservation and Threat

Springboks are listed as Least Concern.

Springboks are usually seen in large herds.

ORYX

The oryx is often called gemsbok, which is its Afrikaans name.

Appearance

The oryx is a striking antelope, with V-shaped horns and a black and white face. Their body is brownish-grey in colour. Their lower legs are white with black upper legs. They have a long black tail and black tips on the ears. Both males and females have horns.

Behaviour

Oryx live in herds which include many females, young males and some non-territorial older bulls. A dominant male leads the herd during breeding times. The size of the herd depends on the amount of food available. In a good season, a herd can include hundreds of animals. Herds move to different areas following the rains, which creates grazing for them.

The oryx is Namibia's national animal.

Oryx can smell when rain is falling some distance away.

The horns of an oryx are powerful weapons. Oryx have been known to kill lions with them.

The horns of the female (left) are usually thinner and longer than those of the male (right).

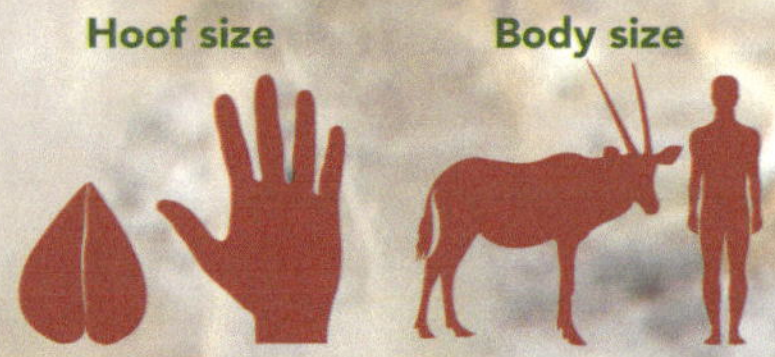

Diet

Oryx eat grasses and leaves. During the dry season, they will eat flowers and thorny shrubs. They also dig up tubers and roots.

QUICK FACTS

Scientific Name: *Oryx gazella*
Conservation Status: Least Threatened
Lifespan: 18-22 years
Predators: lion, hyaena, African wild dog
Habitat: hot, dry areas
Speed: 60 km/h
Horn length: 85 cm

Adaptations

The oryx is adapted to cope in hot, dry environments. It can raise its body temperature to prevent sweating. It also has special capillaries in its nose, which cools down the blood as it travels to the brain. Oryx get most of the water they need from food they eat, so can survive several weeks without drinking water. They will drink if water is available.

In the heat of the day, oryx often rest in the shade of trees.

Raising Calves

Oryx are non-seasonal breeders. However, it is more common for them to give birth from December to April. Female oryx are pregnant for about 9 months, and give birth to a single calf. The calf can run as soon as it is born. After birth, the calf is hidden away for the first few weeks. The mother returns to suckle her calf several times a day. A baby oryx is brown in colour, which keeps it camouflaged in the dry grass. It joins the herd when it gets its adult colouring. Calves are weaned when they are 8 months old.

Dominant males fight to determine who is stronger. These fights can lead to death of one of the animals.

Survival Tactics

The oryx uses its speed to escape predators. When faced with a threat, the herd turns head-on, facing it. If the oryx cannot escape, it will use its horns to defend itself, by going down on its knees, to have maximum effect with the horns.

Oryx usually feed in the early morning and late afternoon.

RED HARTEBEEST

The red hartebeest is a large antelope which has good smell and hearing, but fairly poor sight.

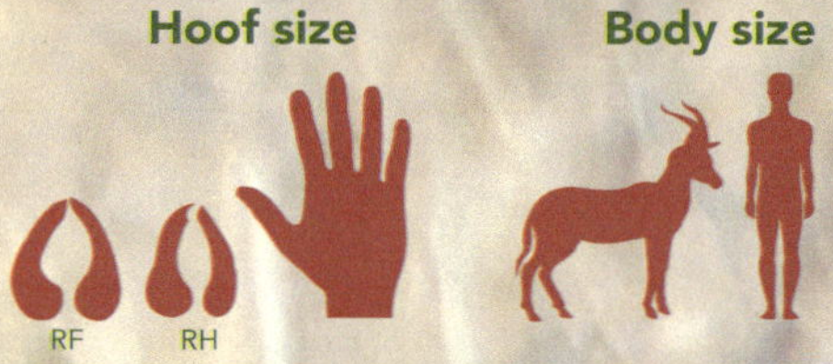

Appearance

Red hartebeest are reddish-brown in colour. They have a black tail and black bands on their legs. Red hartebeest are paler at the top of their back legs and their belly. Their shoulders are humped. They have a straight, long, thin face with a black forehead. A black streak runs down their nose. Both males and females have thick horns, although those of the male are thicker and more robust. The horns meet at the base and curve backwards.

It is thought that the word 'hartebeest' comes from the Dutch words 'hert' (deer) and 'beest' (beast).

Diet

Red hartebeest are mostly grazers, but also eat melons and other forbes in dry areas. They can sniff out good grazing from a long distance away.

Behaviour

They are normally found in small herds made up of territorial males with females and calves. Bachelor males form their own herds. During the rainy season, when there is plenty of food, they sometimes form larger herds. The dominant male marks his territory with dung piles, and often stands on prominent mounds. He does this to keep watch for predators, and also to advertise his position to other hartebeest males. Males fight each other fiercely by locking horns and dragging the other to its knees.

QUICK FACTS

Scientific Name: *Alcelaphus buselaphus*
Conservation Status: Least Concern
Lifespan: 11-19 years
Predators: leopard, lion, cheetah, African wild dog
Habitat: have adapted to most habitats
Speed: 65 km/h
Horn length: 60 cm

Raising Calves

Females are pregnant for 8 months and give birth to a single calf, usually in the summer. The mother leaves the herd to give birth in a sheltered place. The calf is then hidden in dense vegetation. The mother visits her calf to suckle and clean it. It joins the herd when it is strong enough. Males leave the herd when they are 2-3 years old.

Conservation and Threat

The red hartebeest has a conservation status of Least Concern. Their population numbers are increasing. Threats to red hartebeest include habitat destruction, hunting and competing for food.

In Namibia, the largest population of red hartebeest is found on commercial farmland.

© Diganta Sarma

Survival Tactics

When alarmed, red hartebeest act confused, and then run in a zigzag pattern to distract predators. They have good stamina and can run at a speed of 65 km/h.

GREATER KUDU

Appearance

The greater kudu has a brownish-grey coat, a narrow body, long legs, large ears and thin white stripes across its back. Males are much larger than females, and have a beard and dewlap. Only males have long spiral horns, which twist several times like a cork-screw.

The greater kudu is one of the largest antelopes. It can be difficult to spot because it stands very still and is well camouflaged.

A female kudu has no horns or beard. Her nose markings are also not as prominent.

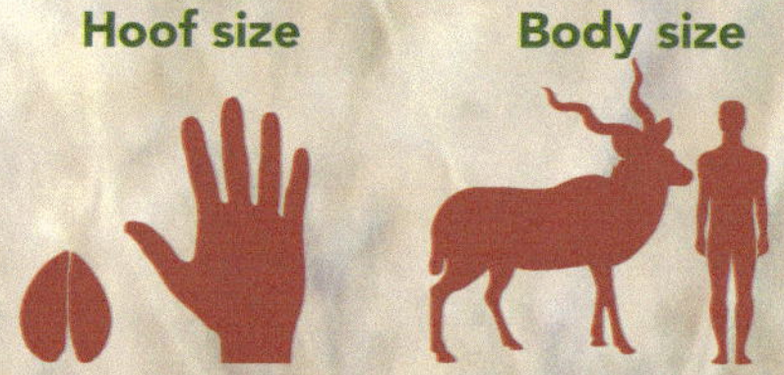

The horns of male kudu do not grow until he is 6-12 months old.

Behaviour

Females form small groups of 3-10 animals. Herds sometimes breakup during the rainy season, when lots of food is available. Male kudu are usually solitary, but sometimes form bachelor groups. Males join female herds during the breeding season. Kudu are mostly active at night, early morning and late afternoon.

Kudu horns can become 187 cm long.

QUICK FACTS

Scientific Name: *Tragelaphus strepsiceros*
Conservation Status: Least Concern
Lifespan: 7-15 years
Predators: lion, leopard, African wild dog
Habitat: woodland, savannah, rocky areas
Speed: 55 km/h
Horn length: 120 cm

Diet

The greater kudu is a browser which eats leaves, shoots and fruit. They are sometimes not dependent on water, but will drink when it is available.

Territorial behaviour

Kudu are not territorial and are not often aggressive, but males sometimes lock horns and push each other to determine who is dominant. Horns often get intertwined and locked during territorial fights and this can lead to the deaths of these animals.

Conservation and Threat

The kudu is listed as Least Concern however humans are converting much of the kudu's natural habitat to farmland. People hunt kudu for their meat, hide and horns. The occasional outbreak of rabies also has a huge impact on populations.

Survival Tactics

Greater kudu spend nearly all their time hidden in thick bush. They have very good hearing, so often hear a predator approaching. When threatened, an adult kudu barks to warn the rest of the herd. Kudu cannot outrun a predator, but can leap higher than 3 metres. They can also move very quietly.

Raising Calves

Greater kudu females are pregnant for 8 months. Calves are usually born between February and March. Females give birth to one calf. The newborn calf is hidden in a number of places in thick bush for 4-5 weeks. During this time, the mother visits her calf several times a day to suckle it. When the calf is 4 months old, it is with its mother at all times. By the age of 6 months, it is independent of its mother but still moves with the group.

The greater kudu can jump higher than any other antelope in Namibia. It easily jumps 3 metres high over very tall fences. Because of this, the kudu is often near the roadside, and can cause damage at night when it jumps in front of vehicles when blinded by the headlights.

BLUE WILDEBEEST

There are two species of wildebeest: the blue wildebeest and the black wildebeest. Only the blue wildebeest, or gnu, is found naturally in Namibia.

Appearance

The blue wildebeest is a large antelope which is grey-black in colour. It has a large head, black mane and pointed beard. The front of its body is heavily built, with slender back legs. Both males and females have horns which curve outwards.

Diet

Blue wildebeest are herbivores which graze. They are dependent on water and need to drink every 2 days.

Behaviour

Blue wildebeest are usually found in herds of 20-30 animals. These herds consist of females and young, and are led by a dominant male. Non-territorial males form bachelor herds. Territorial males defend the females in their territories. They move seasonally for grazing and water, and graze almost all the time. The blue wildebeest is diurnal, mostly active in the early morning and late afternoon. They are noisy creatures and snort a lot while moving.

Males are slightly taller and heavier than females. The horns of males are also larger and heavier than those of a female.

The word 'wildebeest' is an Afrikaans word, which means 'wild beast'.

QUICK FACTS

Scientific Name: *Connochaetes taurinus*
Conservation Status: Least Concern
Lifespan: 17-20 years
Predators: lion, leopard, spotted hyaena, crocodile
Habitat: grassland, savannah, near water
Speed: 70 km/h
Horn length: 35 cm

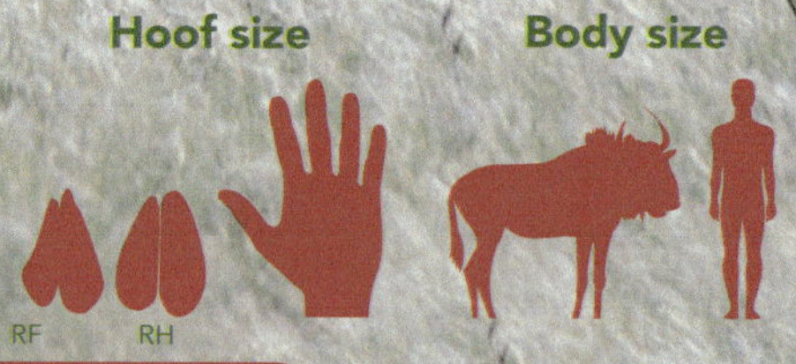

Raising Calves

During the mating season, male wildebeest establish territories and attract females with grunts and posturing behaviour. Females are pregnant for 8,5 months and usually give birth to a single calf. Most females in large populations give birth at the same time to reduce the impact of predation on the young. Calves are born at the beginning of the rainy season, from November to December. They can walk within minutes of birth. Unlike other antelopes, wildebeest females do not leave the herd to give birth. They often give birth in the centre of the herd for protection. Calves are suckled by their mothers for 6 months, although they eat grass after 10 days. Males leave to join a bachelor herd at 1 year old.

Survival Tactics

The main defence of wildebeest is their large herd size and their speed. They keep young animals in the centre, protected by the adults. Wildebeest take turns sleeping, so that some are always on guard. This increases the possibility of spotting a predator in good time. Wildebeest also listen to the alarm calls of other animals.

Wildebeest often group together with Burchell's zebra.

Conservation and Threat

Although they have large numbers, blue wildebeest are declining in areas due to competition with livestock. They carry certain diseases which are a threat to livestock and are therefore not tolerated in areas where livestock farming is popular. Fences around farms prevent these animals from migrating during droughts as they did in earlier times.

BURCHELL'S ZEBRA

The zebra is a member of the horse family. The Burchell's zebra is also known as the plains zebra.

Appearance

Burchell's zebra have horse-like bodies, with short stiff manes. Their tails are tufted at the tip and their coats are striped in black and white. Burchell's zebra have shadow stripes in between their black stripes and a short mane of erect hair. Each zebra has a unique pattern of stripes. Zebra can move quickly for their size, and have teeth built for grinding and cropping grass. Males and females are about the same size, although males are more heavily built.

Zebra are social animals that are found in herds. They graze together and groom each other.

Behaviour

Herds are made up of many family groups consisting of a male, several females and their young. Family members look out for each other. A dominant female leads the group, although the male is the dominant member and is often at the rear of the herd when moving. Males fight for control of the females. Bachelor males live alone or with other bachelors.

What's with the stripes?

It is believed that the stripes make it more difficult for predators to single out an animal because they break up the shape of the animal. The stripes might also help zebra to keep cool.

QUICK FACTS

Scientific Name: *Equus burchellii*
Conservation Status: Least Concern
Lifespan: 20-40 years
Predators: lion, spotted hyaena, African wild dog, leopard, cheetah
Habitat: woodland, open plain, savannah
Speed: 42 km/h

The Burchell's zebra is the closest living relative of the extinct Quagga.

Survival Tactics

Zebra are constantly watching for predators. If the herd is attacked, they run together. Foals run next to their mothers. The dominant male sometimes wards off attackers by biting and kicking. If a zebra is attacked, its family will try to defend it.

Raising Foals

Burchell's zebra are non-seasonal breeders. Females are pregnant for 12-13 months. Single foals are born. Females give birth away from the herd and keep the foal a fair distance from other adults for 2-3 days. Foals are weaned at 11 months. A mother first recognises her own foal by smell, but later by its stripe pattern. A newborn foal can run with its mother within an hour of birth. Foals eat grass after only 3 days. Foals are close to their mothers, but male foals develop relationships with their fathers as well. Male foals leave the herd when they are 3-4 years old and become a threat to the dominant male.

Diet

Zebra are grazers. They prefer short green grass but also eat shrubs, leaves and bark. They need more water than the Hartmann's zebra, so are always near it.

Hartmann's Mountain Zebra

Appearance

The Hartmann's zebra has stripes of black and white. No two zebra have exactly the same pattern. The stripes of the Hartmann's zebra do not cover their white belly. Their stripes are thinner and closer together than those of the Burchell's zebra. Hartmann's zebra also do not have the shadow stripes of the Burchell's zebra. Hartmann's zebra have a dewlap, a mane of short hair and their tail ends in a tuft of hair. They have large, elongated and rounded ears. Males are larger and more robustly built than females.

The Hartmann's mountain zebra is only found in Namibia and Angola. It has small hard hooves which allow it to climb in rocky and mountainous areas.

Behaviour

Hartmann's zebra live in small groups of 7-12 individuals, consisting of one dominant male, several females and their young. They sometimes form temporary groups of up to 30. Bachelors live in separate herds. Hartmann's zebra are diurnal, mostly active in the early morning and late afternoon.

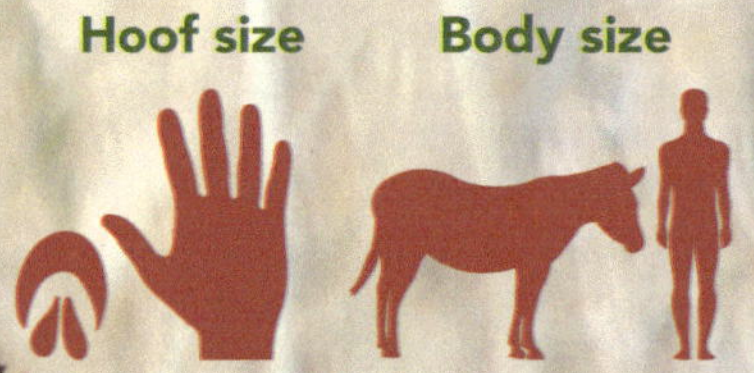

Hartmann's zebra spend up to half of the day feeding.

The heart of the Hartmann's zebra is three times as big as the Burchell's zebra. This allows them to live at higher altitudes.

© Annabelle Venter

QUICK FACTS

Scientific Name: *Equus zebra hartmannae*
Conservation Status: Vulnerable
Lifespan: 20-40 years
Predators: lion, leopard, cheetah, spotted hyaena
Habitat: mountain slopes, grassland
Speed: 32 km/h

Diet

Hartmann's zebra eat tufted grass, bark, leaves, fruit and roots. They are dependent on water. If no surface water is available, they use their sharp hooves to dig for water in dry river beds.

Raising Foals

Females are pregnant for 6 months and give birth to a single foal. Baby zebra are able to stand, walk and suckle shortly after birth. The foal is weaned after 1 year. When a new sibling is born, male young are forced out of the herd.

© Annabelle Venter

Survival Tactics

Hartmann's zebra have very good eyesight with binocular vision in the front. It is suspected that they see in colour. They also have excellent hearing and can hear far away sounds. Hartmann's zebra have a very keen sense of taste and notice slight changes in the quality of their food.

© Annabelle Venter

Conservation and Threat

The Hartmann's zebra is Vulnerable. Their main threat is loss of habitat. They are also hunted for their meat. In the 1930's, this species was hunted to near extinction, with only about 100 individuals left. However, due to many conservation measures, the population has been growing since then and there are estimated to be about 2 700 Hartmann's zebra in the wild today.

GIRAFFE

The giraffe is the tallest living animal. There are 9 subspecies of giraffe. The species found in Namibia is already extinct in Angola.

Appearance

Giraffe have incredibly long necks and cannot be mistaken for any other animal. Their coat has patches ranging in colour and shape depending on subspecies and age. Each giraffe has a unique pattern which can be used to identify individuals. Their tongues are covered in papillae to protect against thorns. Their skin is very thick and they have large hooves. Giraffe have ossicones on the top of their heads which are covered with skin and hair. Males have thicker and slightly longer ossicones which are bald on top while female's are thinner and tufted on top. Males are taller and heavier than females.

Behaviour

Giraffe herd sizes differ from solitary to large mixed herds of males and females which can be spread over a large distance. Individuals or small groups often merge or split from the herd. Giraffe are cathemeral. They sleep only about 4,5 hours a night; either standing up or lying down.

Giraffes can close their nostrils to protect against sand storms and ants.

The heart of a giraffe can weigh up to 11 kg, and has to pump blood all the way up to the brain, which is 2 metres above.

The tongue of a giraffe can be over 50 cm long.

Hoof size **Body size**

Giraffes have no front teeth in their upper jaw, just a 'dental pad'.

QUICK FACTS

Scientific Name: *Giraffa camelopardalis*
Conservation Status: Least Concern
Lifespan: 25 years
Predators: lion, spotted hyaena prey on calves
Habitat: woodland, grassland, savannah
Speed: 50 km/h

Raising Calves

Baby giraffe can be born at any time of year. A female is pregnant for about 15 months, and usually gives birth to a single calf. She gives birth away from the herd and remains with her calf for about a week before returning to the herd. Females give birth standing up and the calf drops to the ground. The new calf's ossicones lie flat on its head when it is born and slowly straighten and fuse to its skull over 3 months. Baby giraffe can stand within 20 minutes of birth. Calves sometimes form a nursery group, with their mothers standing guard. Calves begin browsing by 4 months of age.

Giraffe either spread or bend their forelegs in order to drink because their forelegs are longer than their hind legs.

Survival Tactics

Males solve disputes by swinging their long necks and bashing their ossicones into each other. This is called necking. They defend themselves against predators by kicking with great force.

Diet

Giraffe browse on leaves but also eat fruit, shoots and pods. They seem to prefer acacia leaves. Giraffe can go for several days without water because they get most of the water they need from their food.

Conservation and Threat

Poaching, disease and fragmentation all affect the giraffe population. In poorer African countries, people hunt giraffe for their meat and coat. The tail is also used by many as good luck bracelets, fly whisks and string for sewing beads. Habitat destruction, such as deforestation, also threatens giraffe populations. The subspecies in Namibia is one of the few stable giraffe populations, because of good conservation and management practises. The Giraffe Conservation Foundation (GCF) is working throughout Africa to conserve and protect giraffe in the wild, where numbers of some subspecies are as low as 400 individuals.

WARTHOG

The warthog is a member of the pig family. It has two distinctive pairs of tusks.

Appearance

Warthogs have a large head with two pairs of tusks that stick out of their nose. The tusks are used for digging, fighting and defence. They have a mane of hair down their back, a long tail and short legs.

Warthogs are grey in colour and are mostly hairless, with sparse hair on some parts of their body. Females are usually smaller than males.

Diet

The warthog is an omnivore. Its diet includes grass, roots, berries, bulbs, bark, insects, eggs and carrion. They can go for several days without water if absolutely necessary, but enjoy water when it is available.

Survival Tactics

Although warthog males fight each other, they mostly run away when faced with a predator. They have been known to fight bravely though, even facing down cheetah, leopards and young lions. Their tail is normally upright when fleeing, a sign for others in the sounder to follow. They are not aggressive unless cornered, and will then fight with their tusks causing extensive damage.

Hoof size

Body size

Warthogs are useful for the soil because they aerate and soften it when digging. This allows rain to sink in.

QUICK FACTS

Scientific Name: *Phacochoerus africanus*
Conservation Status: Least Concern
Lifespan: 18 years
Predators: lion, leopard, crocodile, African wild dog, hyaena and cheetah
Habitat: savannah, grassland, woodland
Speed: 48 km/h

Behaviour

Warthogs are not territorial. They live in groups called sounders. Females live with their young and other females or young from the previous litter. Males form bachelor groups. They are diurnal and rest in their burrows at night. Warthogs are powerful diggers. Although they can dig their own burrows, they usually use burrows of other animals. They often reverse into the burrows, with their head facing the opening. They sometimes share burrows with nocturnal animals, such as the hyaena. The hyaena sleeps in the den during the day when the warthog is active and outside, and the warthog sleeps in the den at night when the hyaena is active. Warthogs don't have a layer of fur, so they protect their skin by wallowing in mud to cool down. When it is cold, they huddle together to keep warm.

Warthogs are usually seen singly or in small groups. Their tails are often erect when running.

Conservation and Threat

Warthogs are abundant and there is currently no concern about the survival of this species.

When startled warthogs can run at a speed of 48 km/h.

Raising Piglets

Warthogs are seasonal breeders. Pregnancy lasts for 5-6 months. A litter of 2-8 piglets is born in a burrow. The mother stays with her piglets for the first few weeks of their life. They are completely weaned by about 5 months of age and then begin grazing.

AFRICAN ELEPHANT

Appearance

Elephants are instantly recognisable with their massive grey bodies, large ears, trunk and tusks. They have poor eyesight, but a good sense of smell and excellent hearing. Their tusks are used for digging, foraging and defence. Their trunk is incredibly useful – the elephant can breathe and smell through it, and use it to grab and pick up things. The trunk can hold and squirt water. Male elephants are larger than females.

The African elephant is the largest land mammal. It is bigger than the Asian elephant.

An adult elephant needs to eat 400 kg of food each day, so they spend much of their time travelling great distances in search of food.

The trunk of an elephant has over 100 000 muscles.

Elephants have blood vessels close to the surface of the back of their ears. When an elephant flaps its ears, its blood cools down. Elephants also shower themselves with water or dust to cool down.

QUICK FACTS

Scientific Name: *Loxodonta africana*
Conservation Status: Vulnerable
Lifespan: 65 years
Predators: Lions prey on calves
Habitat: forest, savannah, desert
Speed: 42 km/h
Weight: 6 tons

Behaviour

Elephants are social and form deep family bonds. They live in herds made up of about 10 closely related females and their calves. Herds are led by an older female known as a matriarch. Large herds of up to 100 animals are also seen. Males tend to leave the herd after puberty and either live alone or join other bachelors. Elephants are intelligent and have good memories. They can communicate over long distances with a low frequency rumble that travels through the ground. Elephants show their emotions in different ways, for instance, head-shaking can signify irritation. Charging is often not serious and ends short of the threat, with much trumpeting and kicking up of dust. A silent charge with head down is more serious.

Conservation and Threat

The African elephant is listed as Vulnerable. The two main things that affect the elephant population are habitat loss and poaching. Elephants are poached for their ivory tusks.

Raising Calves

Female elephants are pregnant for 22 months and usually give birth to a single calf. It is the longest pregnancy of all mammals. Calves are cared for by their mothers and other females in the herd. It takes several months before a calf can use its trunk properly, so it suckles through its mouth.

A newborn elephant weighs about 91 kg.

Diet

Elephants eat roots, grasses, fruit and bark.

An elephant's brain is similar to that of a human.

BLACK RHINOCEROS

The black rhino is also called the hook-lipped rhino. Both the black rhino and the white rhino are grey in colour.

Foot size

Body size

Appearance

The black rhino is a large, grey mammal that has two horns. They have a hooked lip and are smaller than the white rhino. Black rhino only have hair on their ears, tail tips and eyelashes. Their horns are made of keratin (just like our fingernails) and grow throughout a rhino's life. Females are slightly smaller than males. They have poor eyesight, but a good sense of smell and hearing.

QUICK FACTS

Scientific Name: *Diceros bicornis*
Conservation Status: Critically Endangered
Lifespan: 35-50 years
Predators: Lions prey on calves
Habitat: prefers thick scrub and bushland, but can exist in semi-desert areas
Speed: 55 km/h
Front horn: 50 cm

Diet

Black rhino browse on leaves, shoots and small branches. Their hooked lip allows them to bite off parts of a plant at a clean angle.

Black rhino have a very thick skin that helps to protect them from thorns.

Behaviour

Black rhino are mostly solitary. Because their eyesight is so poor, rhino rely on scent marking. They mark scent by spraying urine. Black rhino also have scent glands in their skin, and rub their heads on rocks and trees to mark their territories. Black rhino are active day and night, but not during the hottest time of the day. They have strong horns that are used for defence, digging and breaking branches during feeding. Recent evidence has shown that black rhino interact socially while congregating at waterholes.

Conservation and Threat

Black rhino are Critically Endangered. Poaching rhino for their horns is the biggest threat to these animals. Namibia is home to over 90% of the subspecies called the south-western black rhino. Many conservation groups are working towards saving the rhino.

Raising Calves

A male and female rhino usually stay together for only a few days when she is ready to breed. Female rhino are pregnant for 15-17 months. They give birth in a sheltered area. Calves can usually stand within a few hours of being born. They suckle milk from their mothers for up to 1 year. Calves stay with their mother for 2-4 years, before being pushed off when the next calf is born.

Calves weigh 30-45 kg at birth.

Black rhino are more aggressive than white rhino.

WHITE RHINOCEROS

The white rhino is the second largest land mammal, after the African elephant. It is also known as the square-lipped rhino.

Appearance

All rhino are grey in colour. The white rhino has a squared lip, which is very different from the lip of the black rhino. This difference is related to the food they eat – the black rhino is a browser and the white rhino is a grazer. The white rhino is larger than the black rhino, and has a humped back. They only have hair on their ears, tails and eyelashes. They have two horns made of keratin, which grow throughout their life. The front horn is much larger. Females have longer and more slender horns than males, and use them to protect their young.

White rhino can weigh over 2 000 kg and stand as tall as 1,8 metres.

Survival Tactics

White rhino have bad eyesight, so they rely on smell and hearing. Their ears move independently of each other, which helps them to hear more sensitive noises. Their horns are used as defence against predators, and for dominance with other rhino. When calves are pushed off from their mother, they form 'buddy' pairs, so that younger rhino are rarely alone. When threatened, buddies stand back to back, making it very difficult for any predator to attack. This back-to-back position is also adopted by calves after about 1 year old.

QUICK FACTS

Scientific Name: *Ceratotherium simum*
Conservation Status: Near Threatened
Lifespan: 45-50 years
Predators: Lions prey on calves
Habitat: savannah, grassland
Speed: 40 km/h
Front horn: 60 cm

© Christopher L Johnston

Behaviour

Male white rhino are usually solitary. Females sometimes gather in groups (called crashes) of up to 6 animals. Male rhino are only with females during the breeding season. Males are very territorial and defend their territories. Females are not territorial. They occupy home ranges which are much larger than male territories, and may be the same for all females in a population. Male and female territories do overlap. White rhino spend most of their day feeding, resting during the hottest part of day. They take cover by lying in the shade or wallowing in water. If there is no place to wallow, they roll in dust, which protects their skin against the sun and insects.

Conservation and Threat

The white rhino provides one of conservation's greatest success stories. Indiscriminate hunting reduced numbers to less than 50 in the early 1900's. But conservationists managed to save that population, and there are now more than 20 000 in southern Africa. However, poaching has again become a very serious problem, threatening to dramatically reduce population numbers. Many organisations are working to protect this species. Rhino are poached for their horns, which are hard, hair-like growths. The powdered horn is sold as traditional medicine in some Asian countries.

Raising Calves

White rhino females are pregnant for 16 months. They give birth to a single calf. The calf can stand within 1 hour of birth. The calf is suckled for 1 year, although it starts to eat grass at 2 months old. The mother and calf spend 2-4 years together, until the next calf is born. The mother fiercely protects her calf.

White rhino make a variety of noises including huffing and roaring.

© Christopher L Johnston

Calves weigh 40-65 kg at birth.

Diet

White rhino graze. They drink water daily when it is available, but can tolerate longer periods without it.

HIPPOPOTAMUS

Hippopotamus are the third-largest land mammal. They are dangerous and kill more humans than lions.

Appearance

Hippopotamus are large grey mammals. They do not have sweat glands, so keep their bodies cool by staying in water as much as possible. They have flat tails, short legs, and four slightly webbed toes on each foot. Their skin is very thick. They are able to close their nose and ears to block out water, and have membranes which protect their eyes when under water. Their eyes and nostrils are positioned in such a way that they see, hear and breathe, while they are mostly submerged. Hippos have a gland that secretes a red fluid, which helps to protect their skin from the sun.

Hippos eat about 35 kg of grass each evening.

Foot size

Body size

Survival Tactics

The hippo is a very aggressive animal when out of water, or when cornered. They fight using their large teeth and tusks, and can split a boat in two.

QUICK FACTS

Scientific Name: *Hippopotamus amphibius*
Conservation Status: Least Concern
Lifespan: 45 years
Predators: none
Habitat: areas with lots of water
Speed: 23 km/h

Behaviour

Hippos are dependent on water and spend their day in it because their skin needs to stay moist. They go ashore to graze after the sun has set. They live in rivers, lakes and swamps. They cannot swim, so prefer shallower water where they can stand with their head above water. They move through water surprisingly quickly and easily. They can spend up to 5 minutes under water. Hippos are very social, and live in groups of up to 30 animals. Males occupy a territory, while the females and young males form loose groups which move around between male-held territories.

Raising Calves

Females are pregnant for 8 months. The calf is suckled for 18 months. The young calf is kept away from the herd for up to 3 months for its own protection against other adults, before being introduced to the herd.

© Diganta Sarma

Hippos can store food in their stomachs and can go 3 weeks without eating.

Conservation and Threat

Although the hippo is not endangered, it is vulnerable because their numbers are declining. Their main threats are poaching and habitat loss.

The word 'hippopotamus' means 'river horse' or 'water horse'.

Noises

Hippos are very noisy animals, and make sounds such as snorts, grunts and grumbles, that can be heard several kilometres away.

Diet

Hippos are herbivores. They mostly eat grass, but also eat fruit.

ROCK HYRAX

The rock hyrax is commonly called 'dassie', which is its name in Afrikaans. They are often seen in large groups on rocky outcrops.

Appearance

The rock hyrax has a rodent-like appearance, although it is not a member of the rodent family. Rock hyrax have thick fur, which can be greyish-brown in colour, although the colour can vary depending on their environment. Their nose is pointed and they have white eyebrows and whiskers. They have short legs with hoof-like nails. The bottom of their feet has a rubbery texture, which helps them to climb steep rock surfaces and trees.

Diet

Rock hyrax are opportunistic herbivorous. Their diet consists of herbs, grasses, fruit, leaves, roots and bark. They can go for days without drinking water.

Behaviour

Rock hyrax live in large groups of up to 50 animals. They are active in the morning and evening, although this depends heavily on the weather. They sunbathe on rocks to warm up before foraging.

Rock hyrax do not dig burrows. They live in colonies in natural rock crevices. The dominant male defends the group and marks his territory with urine and a gland secretion. They make many noises, such as whistles and shrieks.

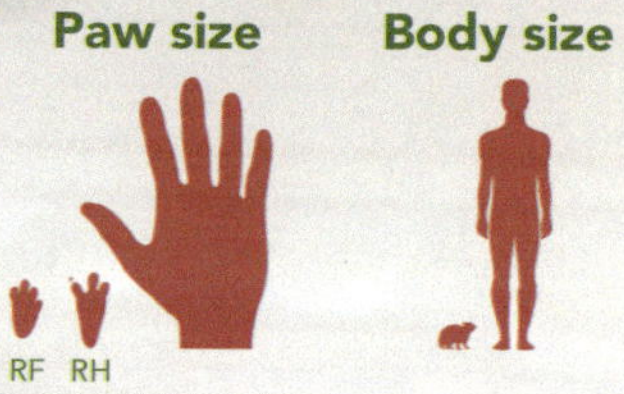

QUICK FACTS

Scientific Name: *Procavia capensis*
Conservation Status: Least Concern
Lifespan: 12 years
Predators: leopard, large raptors, serval, civet, python
Habitat: adaptable, rocky areas
Speed: 30 km/h

The elephant is the closest living relative of the rock hyrax.

Raising Babies

Rock hyrax give birth to 2-3 young after a pregnancy of 6-7 months. Babies are born with their eyes open. Within an hour of birth, rock hyrax babies are able to jump and run. They eat solid food from 2 weeks old and are weaned at 10 weeks.

Conservation and Threat

Rock hyrax are listed as Least Concern. They are fairly widespread with high population numbers in some areas although they are threatened by habitat loss.

Rock hyrax stay in their rocky crevices until the sun is high, and then spend most of the day sun-bathing. Sometimes, if it is cold or rainy, they do not come outside.

Survival Tactics

Rock hyrax have sentries which watch for danger and issue alarm calls when predators approach. Upon hearing the alarm call, members run and hide between rocks. They bite smaller predators with their sharp front incisors to defend themselves.

Rock hyrax have a special claw for grooming.

Rock hyrax can make more than 20 different noises.

GROUND SQUIRREL

The ground squirrel is a member of the rodent family. It mainly lives in dry, semi-desert areas.

Appearance

Ground squirrels have short rough hair with black skin underneath. Their colouring varies from reddish-brown to tawny-yellow. They have a white stripe on both sides of their body. Their underbody, limbs, neck and face are white. They have small ears and a long bushy tail, which is almost as long as their body. They have two dark bands on their tail. Males are larger than females.

Diet

Ground squirrels are omnivores. They eat grass, roots, seeds, fruit, pods, grains, insects, small vertebrates and bird eggs.

Ground squirrels do not wander far from their burrows.

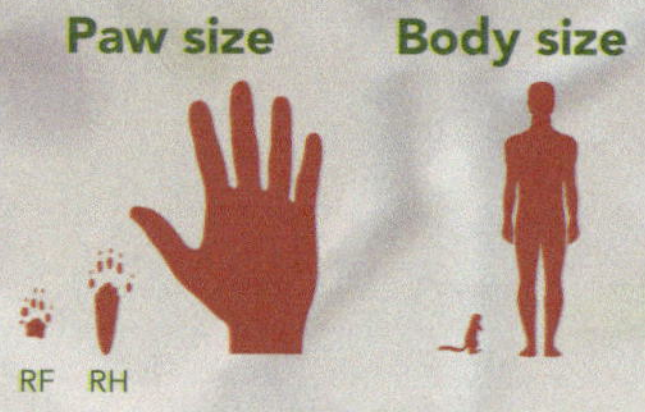

Survival Tactics

If a predator is seen, ground squirrels utter a warning sound, and all members then hide in their burrows.

The burrow of a ground squirrel can have 100 exits.

QUICK FACTS

Scientific Name: *Xerus inauris*
Conservation Status: Least Concern
Lifespan: 6 years
Predators: monitor lizard, black-backed jackal, raptors
Habitat: savannah, grassland, semi-desert

Behaviour

Ground squirrels are very sociable and dig communal burrows, which protect them from weather conditions and predators. They live in colonies of 4-6 females and males. Mature males sometimes form their own groups, called bands. Ground squirrels use their tails as sunshades and are often seen resting with their tails over their heads during the heat of the day. The ground squirrel is diurnal.

After waking, ground squirrels come outside to groom themselves and lie in the sun.

Raising Babies

Ground squirrels have no specific breeding period. Pregnancy lasts for about 6 weeks. 1-3 naked and blind babies are born. Their mother protects them inside the burrow until they are 45 days old. Babies are weaned after 2 months.

Conservation and Threat

Ground squirrels are listed as Least Concern and there is currently no threat to their survival.

A group of ground squirrels is called a dray or scurry.

MEERKAT

Appearance

The meerkat is a small mammal, with a long slender body and black-tipped tail. Meerkats are sandy-brown in colour with darker stripes on their back. The patterns are unique to each meerkat. Their belly and face are lighter in colour. They have a pointed nose, small ears and black patches around the eyes. Meerkats have sharp curved claws for digging burrows and finding prey buried in the sand. The meerkat uses its tail to balance when standing upright and for signalling.

Meerkats are also known as suricates.

Diet

Meerkats are omnivores. They mainly eat insects, but will also eat lizards, snakes, scorpions, spiders, small mammals, eggs and plants. They often forage by digging in soil and overturning rocks.

Behaviour

Meerkats are very social and live in clans of 20-50 members of 3 family groups. Each family group includes a breeding pair and their offspring. They live in large underground tunnels which are lined with grass. They often share burrows with ground squirrels and yellow mongooses.

Clans are very organised. During the day, they sunbathe in the early morning sun, then most go off to forage for food, while others either babysit the young or act as guards. Meerkats are mostly diurnal. If it is raining, they do not come outside. If it is too hot, meerkats return to their burrows to cool down. They are very territorial and fiercely defend their home.

Meerkats have non-retractable claws and four toes on each paw.

Meerkats are immune to certain types of venom.

A group of meerkats is called a mob, gang or clan.

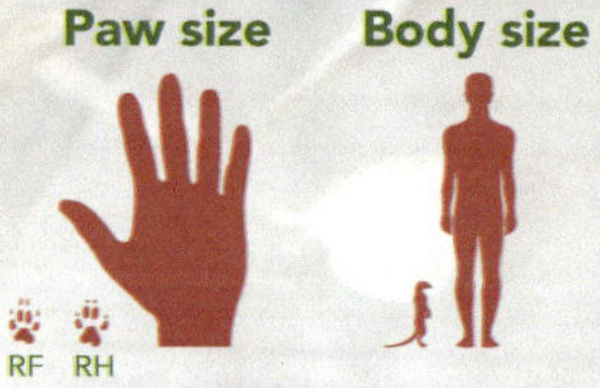

QUICK FACTS

Scientific Name: *Suricata suricatta*
Conservation Status: Least Concern
Lifespan: 5-15 years
Predators: large raptors, black-backed jackal
Habitat: semi-arid areas, savannah

Adaptations

Meerkats are adapted to living in harsh dry environments. The dark patches around their eyes protect them from the glare of the sun. The position of their eyes allows them to see a wide angle, which helps to prevent predators from sneaking up on them. They also have adaptations which help them burrow. Their eyes have a protective membrane that shields them from dirt while digging. Their ears can close tightly to keep dirt out. They have dark skin underneath their fur, and only a thin layer of fur on their bellies. This allows meerkats to warm themselves by lying face up in the sun or when standing.

Survival Tactics

One meerkat is always on sentry duty. It stands on its hind legs, propped up by its tail, and keeps watch. If a predator is spotted, the guard lets out a bark. Upon hearing the signal, all members run to the nearest tunnel. If they cannot get to safety, they stand their ground by arching their back, standing as tall as possible on all four legs with their hair and tail erect, and head lowered. This makes them seem larger. They growl, hiss and rock back and forth. If a predator approaches from the sky, meerkats hide in their burrows. When surprised, adults will protect their offspring by covering them with their own bodies.

Meerkats play an important role in the food chain by providing food for predators and controlling invertebrate populations.

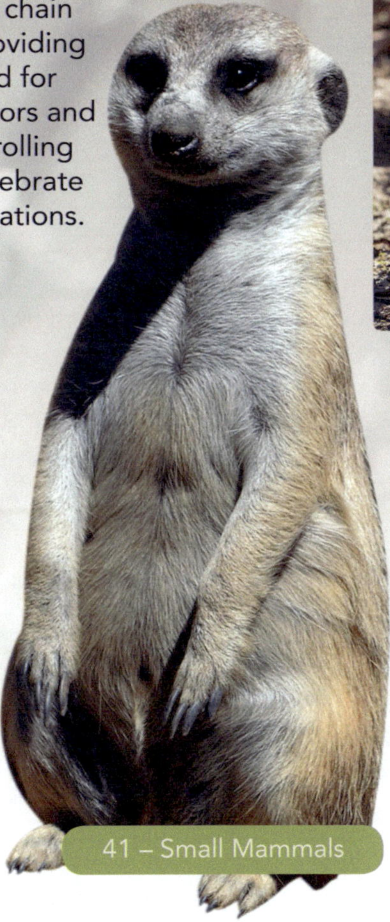

Meerkats are frequently seen standing alert on their hind legs.

Raising Pups

Meerkats give birth throughout the year. Females are pregnant for 22 weeks. They give birth to 2-4 young each year in a burrow. Pups are born with their ears and eyes closed. Fathers and siblings help to raise the young, teaching them about danger and how to forage. Mothers carry young pups by the nape of the neck. Pups start foraging at 1 month old, but are only weaned at 2-4 months old. Non-breeding pack members often act as babysitters.

Conservation and Threat

Meerkats are listed as Least Concern. Although they are widespread and common, populations in certain areas are affected by lack of rain or increase in predators.

BANDED MONGOOSE

Banded mongooses are small mammals. They have non-retractable claws used for digging.

Appearance

Banded mongooses have several dark bands running across their back. They have a large head, pointed face, small flat ears and a pointed tail. Their colouring is greyish-brown, with lighter underparts. They have strong claws which allow them to dig in the soil. Males and females look almost identical.

Survival Tactics

When there is danger, the pack will scatter if there are places to hide. Otherwise, they bunch up, keeping the youngest members safe in the middle. Their large pack size is useful, because they have a better chance of seeing a predator and defending against other animals.

Diet

Banded mongooses are carnivores which eat insects and invertebrates. They occasionally eat rodents, birds, eggs and reptiles, including snakes.

Banded mongooses are curious animals. They inspect any noise or object.

QUICK FACTS

Scientific Name: *Mungos mungo*
Conservation Status: Least Concern
Lifespan: 10-12 years
Predators: raptors, leopard, black-backed jackal
Habitat: savannah, grassland, open forest

Behaviour

Banded mongooses are diurnal and sociable. They live in packs of 10-40 members. They forage for food in the mornings, then rest in the shade during the heat of the day. They usually forage again in the late afternoon. They use their senses of smell, sight and hearing to locate food. They usually forage in groups. Members of the pack help look after young and weak members. They move often, and do not stay in one area for a long time. They live in dens and are not aggressive, but will sometimes fight over food. When packs get too large, both males and females leave to join another pack.

Banded mongooses have been known to remove ticks from warthogs.

© Diganta Sarma

Raising Pups

All females in the group breed and are pregnant for about 2,5 months. Females often give birth within a few days of each other. 2-6 pups are born in a den lined with grass, and are blind at birth. Their eyes open when they are about 10 days old. Pups stay in the den for 1 month, and are always babysat by up to 3 pack members. When the pups leave the den, they start foraging for themselves, although one adult will help and protect them. They are weaned at 3 months old.

Banded mongooses constantly twitter while they forage.

© Diganta Sarma

Males often babysit pups to protect them.

Conservation and Threat

Banded mongooses are rated as Least Concern. There is currently no worry about the survival of this species.

YELLOW MONGOOSE

The yellow mongoose is only found in southern Africa. It is also known as the red meerkat.

Appearance

Yellow mongooses found in the south of Namibia have longer yellow or reddish fur and a white tip on the tail. Those found in more northern areas have shorter tails, greyish hair and a grey tip on the tail. They have a paler belly and chin, and a bushy tail. The head is pointed with large ears. Males and females look the same.

Diet

Yellow mongooses mainly eat insects such as grasshoppers, ants, beetles and spiders. They also eat small rodents, reptiles, frogs, birds, eggs and carrion. Yellow mongooses supplement their diet with nuts and seeds during desperate times.

The yellow mongoose often shares a burrow with ground squirrels and meerkats.

Noises

Yellow mongooses are quiet animals but growl when threatened. They also bark, scream and purr.

With their bushy tail and rounded ears, the yellow mongoose has a fox-like appearance.

Survival Tactics

Their large numbers and early warning systems help to keep them safe from predators. Sentries take turns to keep a lookout around burrows, by standing upright to check for danger.

Paw size **Body size**

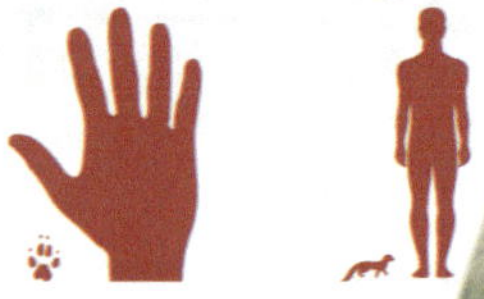

The yellow mongoose is the smallest of all mongooses.

QUICK FACTS

Scientific Name: *Cynictis penicillata*
Conservation Status: Least Concern
Lifespan: 12-13 years
Predators: snakes, jackals, raptors
Habitat: savannah, grassland

Behaviour

Yellow mongooses are mostly diurnal, but are sometimes active in the early evening. They are seen alone or in pairs. They live in extended family groups of 6-8 related animals. A group consists of a breeding pair and their young. Other adults in the group may be related to the breeding pair. They are territorial. Yellow mongooses live in burrows, which have many entrances. Their front feet have long curved claws that are used for digging. They avoid making burrows in hard rocky soil. They spend most of their day foraging and sleep in their burrows at night, or when the weather is unpleasant. Yellow mongooses are often found in urban areas.

Raising Pups

Yellow mongooses breed between July and September. Females are pregnant for about 2 months. Pups are born in a clean burrow. They are weaned after 10 weeks and fully grown by 10 months.

Conservation and Threat

Yellow mongooses are listed as Least Concern because they are widespread and have no major threats.

The yellow mongoose is the only mongoose which brings prey home to its young while in the burrow.

PANGOLIN

Despite its scaly appearance, the pangolin is a mammal, not a reptile. It is also known as the scaly anteater.

Appearance

Pangolins are covered in tough, overlapping scales. The scales are made of keratin – the same protein as your fingernails. Males are bigger than females. Pangolins do not have scales on the underside of their body, which is just soft skin. They have a small head and a long broad tail. They have no teeth, but have a long, muscular, sticky tongue which they use to lap up ants. Pangolins have poor eyesight, but good hearing. They have stout limbs, which are adapted for digging.

Survival Tactics

Pangolin scales are a good defence against predators. When threatened, they roll themselves up into a ball, which protects their undersides. They also puff and lash with their strong tails.

Diet

Pangolins mostly eat ants and some termites. Pangolins use their sticky tongue to lap up the ants. They also have special muscles that close their nostrils and ears to prevent insects from crawling in.

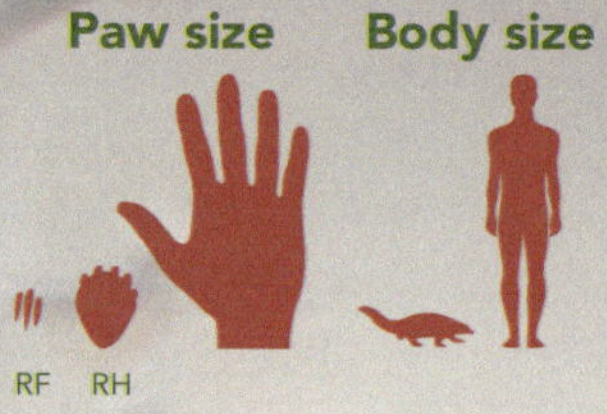

QUICK FACTS

Scientific Name: *Smutsia temminckii*
Conservation Status: Vulnerable
Lifespan: unknown
Predators: honey badger, maybe spotted hyaena
Habitat: savannah, woodland

Behaviour

Pangolins live alone and are very secretive. Because of this, not much is known about them. Some species of pangolin can climb trees, but the species found in Namibia cannot.

Pangolins dig deep burrows, which they sleep in during the day. They dig with their strong front legs and claws, using their tail and back legs for support. As pangolins cannot chew their food, they eat small stones and sand to grind down their meal.

Raising Pups

It is unknown how long pangolin females are pregnant, but it is estimated to be about 5-9 months. They usually give birth to a single offspring. Females raise their young alone. A baby pangolin (called a pangopup) is 15 cm long when born and has hard scales. It is suckled for about 3-4 months. It begins to eat ants at 2 months old. The pangopup then starts to accompany its mother, either riding on her upper back, or sideways across her back. If the mother senses danger, the baby slips under her and is protected when she rolls up her body.

The tongue of a pangolin is longer than its body.

All the pangolin photographs were supplied by REST. They also helped to edit this text, and dispel some of the myths about this mammal.

Conservation and Threat

Pangolins are Vulnerable. They are the most illegally trafficked mammal in the world. The main threat is that they are poached for their scales and meat. Electric fences are also dangerous to pangolins. Many organisations are working towards protecting this species.

CARACAL

The caracal is thought to be the most formidable small cat in Namibia.

Appearance

The caracal is a small cat with a reddish coat. Its underparts are paler and spotted. They are most recognisable by their large black pointed ears, which end in tufts. Caracals have a black line running from the nose to each eye.
They are slender, graceful and have excellent hearing.

Diet

Caracals eat small to medium-sized prey, such as rodents, hares, rock hyrax, birds, lizards and small antelope. They are water independent.

Caracals are elusive and difficult to spot in the wild.

Behaviour

Caracals are nocturnal, and are solitary unless mating or with cubs. They can climb trees but mostly hunt on the ground. They are territorial and mark with scent, by rubbing their head on objects and scratching on trees.

Unlike domestic cats, the pupils of a caracal's eyes contract to circles rather than slits.

© Christpher L. Johnston

QUICK FACTS

Scientific Name: *Caracal caracal*
Conservation Status: Least Concern
Lifespan: 12-17 years
Predators: leopard, black-backed jackal
Habitat: semi-desert, woodland, savannah
Speed: 80 km/h

'Caracal' is from the Turkish word 'karakulak', which means 'black ear'.

Survival Tactics

Caracals stalk their prey. With their speed, lightening fast reflexes, agility and stealth, they are powerful hunters. They can catch prey twice their size and can even catch birds in flight. They are dangerous when cornered and can jump up to 3 metres high.

Caracals have powerful jaws, sharp teeth and curved paws.

Raising Cubs

Caracal females are pregnant for about 2 months. They give birth to 1-5 cubs. The cubs stay with their mother until they are about 1 year old. Their eyes are closed when they are born, but start to open on the first day of their life. They suckle for 10 weeks.

Conservation and Threat

The actual number of caracals in the wild is unknown. They are not considered to be rare or threatened in Namibia, although they are facing habitat loss and are indiscriminately poisoned and hunted by livestock farmers.

© Christopher L. Johnston

Caracals' ears have about 20 muscles.

CHEETAH

Cheetahs are very elegant cats. They are the fastest land mammal.

Appearance

Cheetahs have solid black spots on a tawny-golden coat. They are built for speed, with a small head, large chest, slender waist and slim legs. They have a white throat and belly. Cheetahs have a long spotted muscular tail, which ends in black stripes, and helps to keep control and balance when running.

Diet

Cheetahs are carnivores who mostly feed on small-sized antelope. When several cheetahs hunt together, they can take down larger animals such as wildebeest and zebra. They can last 4-10 days without water.

Behaviour

Cheetahs rely on tall grasses for camouflage when hunting. They are mostly diurnal, usually hunting in the late morning or early evening. Females are solitary, except when raising cubs. Brothers sometimes form a group. Cheetahs cannot roar like other big cats, but can purr, hiss, whine and growl.

Cheetahs have 'tear drop' lines, which protect their eyes from the glare of the sun.

The word 'cheetah' means 'spotted one' in Hindi.

QUICK FACTS

Scientific Name: *Acinonyx jubatus*
Conservation Status: Endangered
Lifespan: 16 years
Predators: eagles, hyaena, lion
Habitat: open grassland, savannah
Speed: 110 km/h

Raising Cubs

Cubs are born all year round, after a pregnancy of 3 months. A litter normally consists of 2-5 cubs. Cheetah cubs are born with a mantle of fur, which runs from the back of the neck. This helps the cubs to disguise themselves in grass, and makes them look slightly like a honey badger. The mantle starts to disappear after 3 months. Cubs have grey coats which become golden and spotted after 4 months. Mothers move their cubs to new hiding places every few days to keep them safe from predators. When they are 6 months old, she will fetch live prey so that her cubs can practise hunting and killing.

Survival Tactics

Cheetahs usually hunt by hiding in tall grass and sneaking up on their prey, often choosing an animal that has strayed from the herd. The cheetah eats immediately, before any scavengers can take its meal.

Cheetahs have non-retractable claws, which help to grip the ground when running.

When running, cheetahs use their tail to help them steer, like the rudder of a boat.

Conservation and Threat

It is estimated that there are less than 10 000 cheetah left in the wild. It is Africa's most endangered cat. The biggest threat to the cheetah is habitat loss. The cheetah is also poached for its coat. Because cheetahs hunt domestic animals, farmers sometimes poison or hunt them.

LEOPARD

Leopards are closely related to lions, tigers and jaguars. They are well camouflaged and very difficult to spot.

Appearance

Leopards generally have a creamy golden fur with dark, irregular spots called rosettes. Leopards have sleek, powerful bodies. They can swim, climb, run and jump. Their long tail helps with balance and allows them to take sharp turns.

Diet

Leopards eat a variety of food, taking advantage of what is available. They eat small and medium-sized mammals, as well as birds, rodents, fish and fruit.

Behaviour

Leopards are nocturnal, resting in trees or bushes during the day. They are solitary and have large territories. The territories of males and females can overlap. Leopards mark their territory with urine and claw marks on trees. Leopards often leave their prey in a tree and return to feed on it when they are hungry.

QUICK FACTS

Scientific Name: *Panthera pardus*
Conservation Status: Near Threatened
Lifespan: 21 years
Predators: lion, eagles, spotted hyaena, African wild dog prey on cubs
Habitat: wide range, adaptable
Speed: 45 km/h

Leopards growl when angry and purr when happy. They communicate and signify territory with a loud saw-like growl.

Raising Cubs

Female leopards are pregnant for 3-4 months. They give birth to 1-6 cubs, who are blind at birth. The mother stays in the same territory as her cubs until they are old enough to travel with her. Cubs are kept hidden for 8 weeks, and are suckled for 3 months. Cubs stay with their mother for about 2 years.

Survival Tactics

Leopards are adaptable and can live in many different habitats. They have long canines and powerful muscles in their necks, jaws and shoulders. This allows them to kill medium-sized antelope. A leopard does not have the speed of the cheetah, so relies on stealth to catch prey. They do this by slinking quietly through the grass or lying in ambush until they are close enough to pounce. Leopards hunt at night. A leopard depends on hearing and eyesight when hunting.

Conservation and Threat

Leopards are Near Threatened. They are hunted for their coats, for sport and by farmers, whose livestock leopards sometimes prey on.

Leopards are often seen in trees. The main reason for this is to protect themselves and their food.

LION

Appearance

The lion is a large cat, whose colour varies from yellow to reddish-brown, with a lighter stomach. Most males have a thick mane of brown/black hair on the head and neck. Their tail ends in a tuft of hair.

The lion is the biggest cat in Namibia.

An adult male's roar can be heard up to 8 km away.

Behaviour

Unlike other cats, lions live in a group, called a pride. A pride consists of family units of about 3 males, several females and their young. All these females are related. Lions are lazy, spending 16-20 hours a day sleeping or resting. The rest of the day is spent hunting and protecting their territory. They keep in touch with each other by roaring. Not all lions live in prides. Young males often spend years alone until they are strong enough to take over a pride of their own. Lions are very affectionate towards each other while resting.

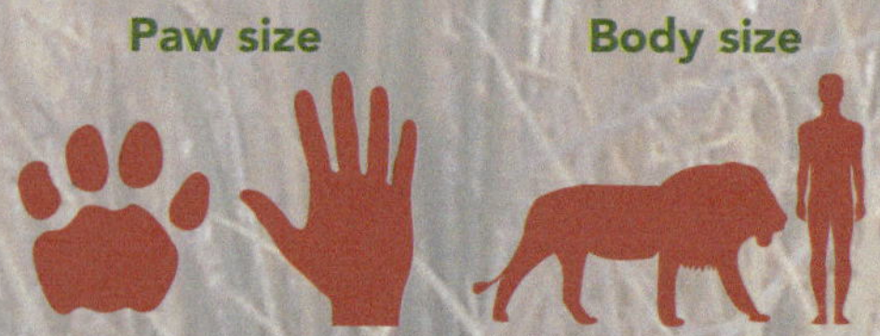

Lionesses do most of the hunting, while males patrol and protect the pride.

QUICK FACTS

Scientific Name: *Panthera leo*
Conservation Status: Vulnerable
Lifespan: 15-30 years
Predators: hyaena, leopard; black-backed jackal prey on cubs
Habitat: grassland, savannah
Speed: 81 km/h

Survival Tactics

Lions prefer to hunt at night, although they sometimes hunt during the day. They hunt in a team, which increases their success. When they spot a herd, they form a semi-circle around it and slowly creep towards the prey. When they are close enough to strike, they leap into the herd and startle the animals. Then they knock one down and kill it by biting the neck or throat. Hunts are successful about half the time.

Raising Cubs

Females in a pride usually give birth to 2-3 cubs at the same time, and the cubs are raised together. Mothers nurture their young unless food is scarce. Cubs are capable hunters after 2 years. Cubs are vulnerable to predators such as spotted hyaenas and leopards. If another male lion takes over a pride, he will kill all cubs. Young males are forced out of the pride at 2 years, and usually form bachelor groups. They are fully grown at about 5 years old.

Conservation and Threat

The lion is listed as Vulnerable. Drought, disease, hunting and human encroachment are affecting the lion population. Some farmers also poison lions who eat their livestock.

Diet

Lions eat a wide variety of prey, from large to small, including wildebeest, warthog, impala, zebra and giraffe.
If water is available, they will drink daily, but can survive 3-4 days without water.

The mouth of a lion can open as wide as 30 cm, allowing it to suffocate its prey.

A male lion needs to eat 7 kg of meat each day.

AFRICAN WILD DOG

African wild dogs depend on each other for survival. Everything they do is for the benefit of the pack.

Appearance

The African wild dog has a beautiful coat of black, white, grey and yellow-brown blotches. Each individual has a unique pattern, much like our finger prints, which is used to identify members of the pack. Pups are a dark brown/black colour when they are born. Their patterns emerge as they get older. The hair of the African wild dog is shorter on the legs and body, but longer on the neck. They have a dark face and stripe up the forehead. Ears are large and rounded. The tail is bushy with a white tip. It has only four toes on the front feet.

Survival Tactics

African wild dogs hunt daily, usually in the early morning or late evening. They hunt in packs, and chase their prey over long distances. They run fast and about 80% of their hunts are successful. They hunt together, which allows them to take down larger animals. Pups are allowed to eat first. There is no aggression between members when feeding.

Adaptations

Their large ears can pick up sounds from a long distance away, and this helps pack members keep in contact with each other. The African wild dog has very powerful jaws, which allows it to grab and hang on to its prey. Its coat gives excellent camouflage.

Diet

African wild dogs are carnivores who eat small to medium-sized antelope.

Paw size

Body size

Their Latin name *'Lycaon pictus'* means 'painted wolf'.

QUICK FACTS

Scientific Name: *Lycaon pictus*
Conservation Status: Endangered
Lifespan: 10-12 years
Predators: lion, spotted hyaena
Habitat: savannah, woodland
Speed: 56 km/h

African wild dogs are the most successful pack hunters in Namibia.

Behaviour

African wild dogs are very sociable, with a strict hierarchy and elaborate displays of 'hoots' and facial licking, all of which help strengthen the bond between the pack.

Pack numbers range from 5-27 members. Each pack has a dominant breeding pair. When they have pups, they live in a den, often already dug out by other animals (such as the aardvark), and just enlarge it. At other times they roam, travelling long distances in search of prey or new territory, sometimes as far as 72 km in one night. The pack looks after its members, sharing food and helping weaker or ill members.

African wild dogs are highly intelligent and communicate through sound and body language.

Raising Pups

There is only one breeding pair per pack. A pregnant female usually gives birth to 4-12 pups. The entire pack cares for the pups. Pups are born in a shelter or den, and stay there for 3 months with their mother. When the pack returns from hunting, they regurgitate food for her. As pups get older, the pack helps with babysitting. Pups are good hunters from about 1 year old. When they are 2 years old, females leave the pack to join a new one, but males stay with their original pack. The mortality rate in pups is very high.

Conservation and Threat

The biggest threat to African wild dogs is direct persecution by farmers who shot and poison them to stop them eating livestock. Other major threats include habitat and prey loss (prey being poached by humans) as well as road mortalities. African wild dogs are also prone to diseases such as canine distemper and rabies, which they can catch from domestic dogs that they come into contact with. Unfortunately, the social behaviours of the African wild dog which help bond the pack, such as face licking, also spread disease through the pack and can lead to entire packs dying out in an area. Competition and predation with large predators (lion and spotted hyaena) also threatens the survival of this species and limits where they can live successfully.

© Luke Street

BAT-EARED FOX

The bat-eared fox gets its name from its bat-wing shaped ears.

Appearance

Bat-eared foxes are well camouflaged with their sandy-grey to deep-honey coloured coat. They have lighter fur on their belly, which is darker around their eyes, muzzle, tip of tail and back of ears. They have a long bushy tail.
Bat-eared foxes have teeth which are pointed and specialised for crushing, rather than chewing, their food. Their claws are adapted for digging.

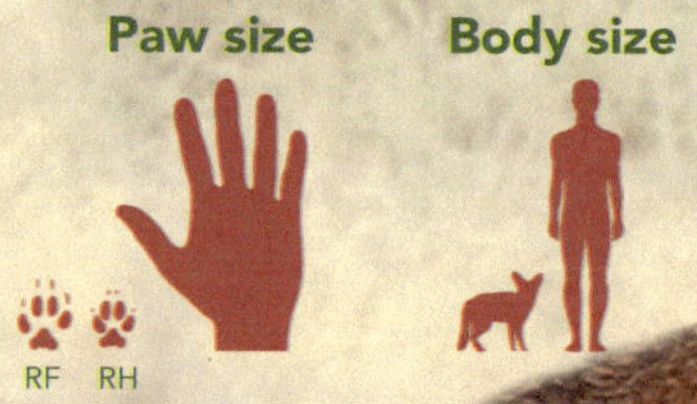

Survival Tactics

The bat-eared fox relies on its speed, hearing and dodging ability to escape predators. Staying in large groups also increases the chance of seeing a predator in good time.

Hunting Techniques

Bat-eared foxes usually forage in groups. Prey is located by hearing, rather than by sight or smell. Their foraging times usually coincide with termite availability and activity. For instance, in summer they begin foraging at sunset, whereas in winter they forage earlier in the day.

The species name '*megalotis*' comes from the Greek words 'mega' and 'otus' meaning 'large eared'.

© Annabelle Venter

The large ears of the bat-eared fox give it excellent hearing. They are also full of blood vessels that release heat and keep the fox cool.

QUICK FACTS

Scientific Name: *Octocyon megalotis*
Conservation Status: Least Concern
Lifespan: 6-13 years
Predators: lion, leopard, cheetah, spotted hyaena, African wild dog
Habitat: Desert/semi-desert, open plains, grassland, savannah
Speed: 48 km/h

Behaviour

Bat-eared foxes live in packs of 2-15 individuals, which consist of a breeding pair and their pups. Males and females mate for life. They live in several dens, each with many entrances, tunnels and chambers, which they dig themselves. They are also known to use old termite mounds as dens. Dens are used for sleeping, giving birth and raising pups. Bat-eared foxes are normally nocturnal, however this depends heavily on the weather and food availability. They are not strictly territorial, and ranges of different packs overlap where food is abundant. Members of unrelated groups often play, forage and rest together. Social grooming also occurs.

Conservation and Threat

Bat-eared foxes are not currently a threatened species, but they are losing their habitat and are hunted for their coat. Other things affecting their population include diseases, such as rabies. They often fall prey to poisons put out for other predators by unscrupulous livestock farmers.

Raising Cubs

Bat-eared foxes usually mate for life. Females are pregnant for 2 months, and give birth to 1-6 cubs. They are suckled for 14 weeks. After this, the father looks after the cubs, including grooming, defending and carrying them between dens. Cubs leave the family when they are about 6 months old.

The ears of the bat-eared fox can grow up to 14 cm long.

Bat-eared foxes play an important role in termite control. It is estimated that an adult eats up to 1 million termites every year.

© Annabelle Venter

Diet

Bat-eared foxes are insectivorous, mainly eating termites and ants. They also eat beetles, grasshoppers, scorpions, spiders, millipedes, rodents, lizards, fruit and eggs. They are not dependent on water.

Bat-eared foxes have special muscles in their jaw, which allow them to chew rapidly, at 5 times per second.

BLACK-BACKED JACKAL

The black-backed jackal has dog-like features and a bushy tail.

Appearance

Black-backed jackals are reddish-brown/tan in colour. They have a distinctive black band running down their back and a bushy tail. They have light fur inside their legs and on their belly. Black-backed jackals have long legs and can run for a long time. Males are slightly larger and heavier than females.

Behaviour

Black-backed jackals live alone or in pairs. Males and females mate for life and share a territory which they both mark and defend. The pair is closely bonded and spend most of their time together. They share activities such as scent marking, hunting and eating. Even the smallest scrap is shared equally. They are sometimes found in small packs. They are mainly nocturnal, but in game reserves are often active during the day. They live in dens which can be up to 2 metres deep.

Survival Tactics

The black-backed jackal is an opportunistic and resourceful animal. They scavenge food which is easy to get, but also hunt their own prey. They follow large carnivores who have made a kill, keeping their distance until the big predators have had their fill. Then they dart in to eat before the vultures arrive.

This jackal is often seen on roadsides in commercial farming areas and frequently in Etosha National Park.

QUICK FACTS

Scientific Name: *Canis mesomelas*
Conservation Status: Least Concern
Lifespan: 10-14 years
Predators: lion, leopard, spotted hyaena
Habitat: wide range, very adaptable
Speed: 12-16 km/h

The black-backed jackal is also known as the silver-backed jackal.

Conservation and Threat

The black-backed jackal is currently listed as Least Concern. There are, however, threats to this species because they occasionally kill the young of domestic animals, such as goats and sheep. Therefore they are in constant conflict with farmers, who poison or trap them in order to protect their livestock.

Diet

Black-backed jackals are omnivores. Their diet includes reptiles, birds, snakes, eggs, invertebrates, small mammals, young antelope, fruit, berries and grass.

This jackal is waiting for the cheetah and her cubs to finish their meal.

Raising Pups

1-9 pups are born in an underground den between July and October. The pups are blind until they are about 10 days old. For the first 3 weeks, they are constantly cared for by their parents, and suckle milk from their mothers. They spend most of their time in the burrow until they are 7 weeks old. They can hunt by themselves at the age of 6 months. Sometimes pups stay with their parents and help to raise younger brothers and sisters.

Jackals are very noisy and communicate with whines, howls, yelps, woofs and growls.

BROWN HYAENA

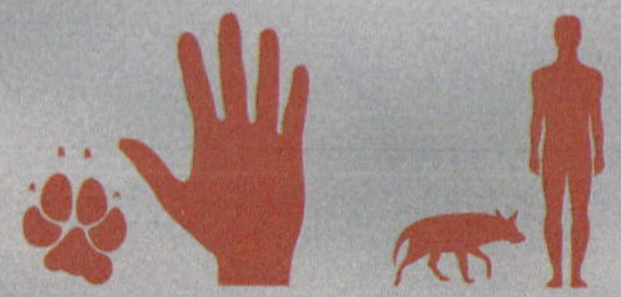

The brown hyaena is only found in southern Africa.

Appearance

The brown hyaena has the typical hunched-backed hyaena shape, with longer, thicker front legs and shorter back legs. Their back feet are much smaller than their front feet. The brown hyaena has a long, shaggy dark brown coat. It has brown and white stripes on its legs. Adults have cream-coloured shoulders and neck, with longer hairs which form a mane. Males and females look the same.

Behaviour

Brown hyaenas forage alone but live as members of a clan. One third of males are usually nomads. Females can be nomadic, but only when they are not raising cubs. Clans consist of up to 15 individuals, and are led by a dominant male and female. A clan defends a territory, and all members help with raising cubs. Brown hyaena are nocturnal, and rest under cover during the day to avoid the heat. Food that is not eaten immediately or brought back to the cubs, is buried. Brown hyaenas have latrine sites for defecation throughout their territory. Cubs leave the den to defecate.

Brown hyaenas play a valuable role in controlling the spread of disease and parasites, by eating dead animals before the meat rots.

The ears of the brown hyaena are larger and more erect than those of the spotted hyaena.

Brown hyaenas have a good sense of smell.

QUICK FACTS

Scientific Name: *Hyaena brunnea*
Conservation Status: Near Threatened
Lifespan: 12 years
Predators: lion, spotted hyaena
Habitat: savannah, grassland, desert
Speed: 40 km/h

Brown hyaena travel 20-35 km each night in search of food.

Conservation and Threat

Brown hyaenas are Near Threatened. Their low population numbers are probably because of the dry areas they inhabit, as well as persecution from livestock farmers. Like most hyaena species, the brown hyaena is often the victim of myths. There is a false belief that brown hyaenas hunt livestock, and because of this commercial farmers often poison, trap and hunt them. But in reality, brown hyaenas rarely kill livestock.

Raising Cubs

Brown hyaena females are pregnant for about 3 months and give birth to 1-5 cubs. Mothers give birth in separate dens. When cubs are 4 months old, they are moved to a communal den. All mothers in the clan suckle all cubs, and all clan members bring food to the den for young cubs. Cubs are born deaf, and with their eyes closed. From 3 months old, meat is brought back to the cubs. Cubs are weaned at about 1 year old. They spend most of their time close to the den until they are 15 months old, when they start foraging on their own.

© Annabelle Venter

Brown hyaenas have incredibly strong teeth and jaws which can crush large bones easily.

Diet

The brown hyaena is an opportunistic feeder, which mostly scavenges. They do not usually hunt live prey, but will catch small mammals or birds if the opportunity arises. During the dry season, they also eat melons, which have a high water content. This helps them to survive in the dryer areas of Namibia.

SPOTTED HYAENA

The spotted hyaena is also known as the laughing hyaena.

Appearance

The spotted hyaena is the largest species of hyaena. They have the most powerful jaws of all predators. Their front legs are longer than their back legs, which makes their back look humped. They have non-retractable claws and a short, brown tail. They have a short rough coat which is sandy, ginger or grey in colour, and covered in spots. The colour of the fur changes with age. Their ears are rounded and they have a short erect mane of hair along their back. They have an excellent sense of smell, good hearing and sharp eyesight at night. They have larger brains than other hyaena species.

Behaviour

Spotted hyaenas are very intelligent animals. They live in clans of 4-20 adults. Each clan occupies and defends a territory and is led by a dominant female. Females are larger and more aggressive than males. Clan members compete with each other over food, but work together as a team when hunting. When they are apart, they keep in contact by making loud sounds. Spotted hyaenas live in dens. In the heat of the day, they often cool off in shallow water or under bushes. They are nocturnal but can be active in the day.

© Luke Street

Paw size

Body size

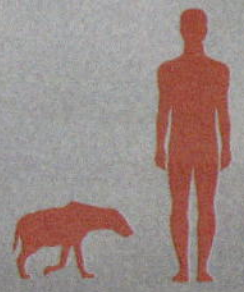

QUICK FACTS

Scientific Name: *Crocuta crocuta*
Conservation Status: Least Concern
Lifespan: 25-40 years
Predators: lion
Habitat: savannah, semi-desert, grassland, woodland
Speed: 60 km/h

Although hyaenas look similar to dogs, they are more closely related to cats.

© Luke Street

Noises

Spotted hyaena can make over 11 different sounds. They whoop, yell, growl, whine, squeal and laugh.

Raising Cubs

The spotted hyaena is a non-seasonal breeder. A female is pregnant for about 4 months and gives birth to 1-4 cubs. Cubs have teeth when they are born, and their eyes are open. Within minutes of birth, they act aggressively. Cubs are born with dark brown hair.

If several hyaena have cubs, the young will be moved to a nursery den. A female hyaena only feeds her own cubs and is very protective over them. Cubs develop their adult colouring at 2-3 months old. Males do not help in raising the cubs. They are weaned at 12-16 months old, but eat meat from 9 months. Female cubs stay with their clan. At least half of all spotted hyaena cubs die before they become adults.

Hyaenas can hear predators from 10 km away.

Survival Tactics

The spotted hyaena is often portrayed as a scavenger, and although it does scavenge, it hunts more than half of its food. They hunt in groups – the bigger the group, the bigger the prey.

They often pick out young or weak animals as their target. Spotted hyaena track their prey by sight, hearing and smell. They have good stamina and speed.

Conservation and Threat

The spotted hyaena is not currently threatened. However, they are experiencing decline in some areas due to poaching, habitat loss and drought. The spotted hyaena is also shot, poisoned, trapped and hunted by farmers when hyaenas kill their livestock.

Diet

Spotted hyaenas eat quickly. Their diet includes carrion, antelope, warthogs, jackals, eggs and possibly pangolins. They do not need to drink much water.

Females are bigger than males.

© Luke Street

CHACMA BABOON

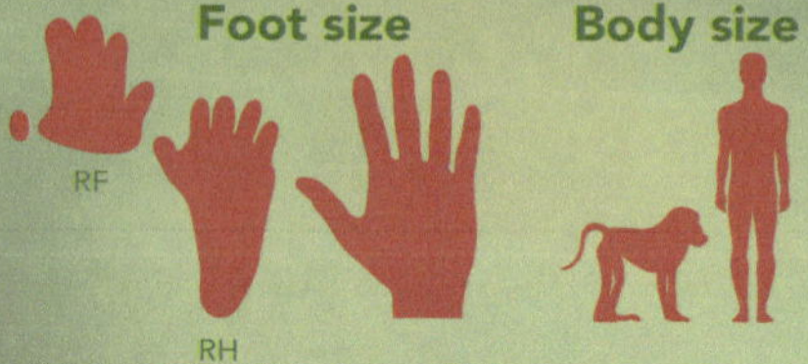

Baboons are fairly common and are often seen foraging by the side of the road.

Appearance

Baboons have a scraggly-looking coat which is brownish-grey in colour, and a paler chest. They have a long tail and a dog-like muzzle. They have a sloping face with large canine teeth and strong jaws. Males have manes and are larger than females.

The chacma baboon is the biggest member of the monkey family in Namibia.

© Christopher L Johnston

Diet

The chacma baboon is an omnivore. They eat fruit, all tree parts, roots, berries, eggs, insects, mushrooms, small mammals, birds and lizards. They are dependent on water.

The chacma baboon is also known as the cape baboon.

Baboons sometimes have conflict with humans. Troops can become a menace by turning over rubbish bins and entering houses in search of food. Baboons are dangerous when threatened.

QUICK FACTS

Scientific Name: *Papio ursinus*
Conservation Status: Least Concern
Lifespan: 30-40 years
Predators: leopard, lion, cheetah, python
Habitat: savannah, rocky areas

Behaviour

Baboons are very social and live in large troops of 10-200 individuals. Troops consist of adult males, females and their offspring. Males and unrelated females form strong bonds. Baboons are diurnal. They sleep in large groups on cliffs or in trees to avoid predators. They forage on the ground or in trees during the day. Males are dominant and fight for their position. A female's position is inherited from her mother. Grooming is important for baboons as it strengthens relationships and indicates hierarchy.

Conservation and Threat

Baboons are listed as Least Concern. They are widespread and there is currently no worry about their survival. However, conflict with humans and drought have negative effects on populations.

Raising Infants

Baboons have no strict breeding season. Females are pregnant for 6 months and give birth to a single infant. Infants cling to their mother's underside for the first few weeks, and start riding jockey-style when they are about 7 weeks old. They can climb trees from 4 months old. They are weaned at 6 months. Males are protective of their young and will sometimes babysit. Male offspring leave the troop when they are 2 years old.

The tail of a baboon does not grip to climb, so they spend more time on the ground than other primate species.

© Christopher L Johnston

Baboons make a wide range of calls including grunts and screeches.

Survival Tactics

Baboons have cheek pouches, in which they can store food. Although they are dependent on water, they can survive for about 20 days without it, by eating food with a high water content.
Their large numbers help to keep baboons safe from predators. Males are usually on sentry duty. If a predator is spotted, a warning bark is let out to alert the troop and members then scurry to safety. Males sometimes gang up on a predator.

VERVET MONKEY

In Namibia, vervet monkeys are only found in the far northern and southern parts.

Appearance

Vervet monkeys have a white-greyish coat. Their face, hands and tail tip are black. They have a white band above their eyes, and whitish underparts. They have a long stiff tail. Vervet monkeys move on all four limbs, both on the ground and in trees. They can also leap from tree to tree. Males are larger than females and have a blue scrotum.

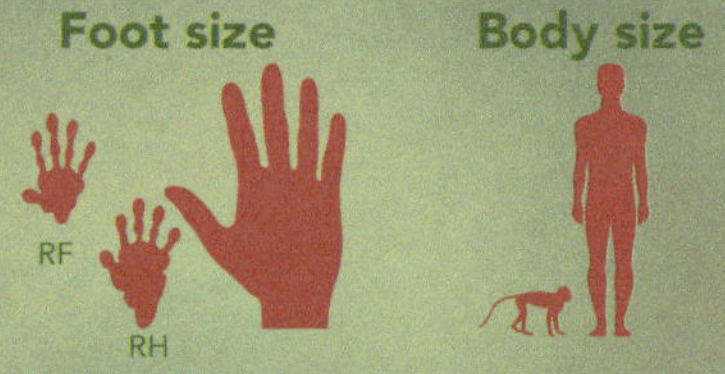

Vervet monkeys can make up to 36 sounds.

Behaviour

Vervet moneys are diurnal, spending their days foraging and sleeping at night. They live in troops of 15-50, led by a dominant male. They prefer to be in trees, although also search for food and water on the ground. Groups of 2-3 monkeys sleep together at night in rock crevices or tall trees. Males do sentry duty and defend against predators. Troop members are very caring towards other members, and are often seen grooming each other.

Survival Tactics

Vervet monkey avoid many predators by staying close to trees. They live in large troops, so there is more chance of spotting a predator in good time. They defend themselves and their young. They scream and whistle to alert other troop members of danger.

QUICK FACTS

Scientific Name: *Ceropithecus aethiops*
Conservation Status: Least Concern
Lifespan: 24 years
Predators: leopard, python, baboon, large raptors
Habitat: savannah, woodland

© Christopher L Johnston

Conservation and Threat

The vervet monkey is listed as Least Concern.

Diet

Vervet monkeys are omnivores, although they mostly eat fruit, flowers, seeds and leaves. They will also eat spiders, grubs and locusts when hungry.

Vervet monkeys groom each other, removing dirt and bugs from each other.

Raising Infants

Females are pregnant for 6-7 months and give birth to a single infant. The baby clings to its mother stomach for the first week of its life. They are born with pink faces and black fur. They start to look more like their parents when they are 4 months old. They suckle for 1 year, but start to forage at 4 months. Fully grown males leave to join other groups, but females remain with their mother's troop. Females have been known to adopt orphaned infants from other troops. If her baby dies, the mother will sometimes carry its body around with her for days, mourning the loss of her child.

Vervet monkeys that live near humans can become pests by stealing food or raiding crops. They are good climbers and often get away when chased.

© Christopher L Johnston

Martial's Mammals is also available as an interactive book through the iBook store.

Made in the USA
Monee, IL
05 November 2026

67948874R00045